MINNEAPOLIS-MOLINE U-SERIES

PHOTO ARCHIVE

MINNEAPOLIS-MOLINE U-SERIES

PHOTO ARCHIVE

Photographs from the
Minneapolis-Moline Company Records

Edited with introduction by
P. A. Letourneau

Iconografix
Photo Archive Series

Iconografix
P.O. Box 18433
Minneapolis, Minnesota 55418 USA

Library of Congress Card Number 93-79371
ISBN 1-882256-07-7

93 94 95 96 97 98 99 5 4 3 2 1

Cover and book design by Lou Gordon, Osceola, Wisconsin
Digital imaging by Pixelperfect, Madison, Wisconsin

Printed in the United States of America

PREFACE

The histories of machines and mechanical gadgets are contained in the books, journals, correspondence and personal papers stored in libraries and archives throughout the world. Written in tens of languages, covering thousands of subjects, the stories are recorded in millions of words.

Words are powerful. Yet, the impact of a single image, a photograph or an illustration, often relates more than dozens of pages of text. Fortunately, many of the libraries and archives that house the words also preserve the images.

In the Photo Archive Series, Iconografix reproduces photographs and illustrations selected from public and private collections. The images are chosen to tell a story... to capture the character of their subject. Reproduced as found, they are accompanied by the captions made available by the archive.

The Iconografix Photo Archive Series is dedicated to young and old alike, the enthusiast, the collector and anyone who, like us, is fascinated by "things" mechanical.

ACKNOWLEDGMENTS

The majority of photographs which appear in this book are part of the Minneapolis-Moline Company Records, and were made available by the Minnesota Historical Society. We are most grateful to the society, and sincerely appreciate the cooperation of its staff. Supplemental photographs were provided by the editor.

MINNEAPOLIS-MOLINE

INTRODUCTION

Minneapolis-Moline Power Implement Company was organized in April 1929, through the merger of Minneapolis Steel and Machinery Company (MS&M), Moline Implement Company (MIC), and Minneapolis Threshing Machine Company (MTM). At the time of the merger, both MS&M and MTM built a line of tractors. However, the new company suspended production of the MTM "Minneapolis" line and carried-on with the MS&M "Twin City" line.

In 1931, Minneapolis-Moline introduced the Universal Model MT, the company's first row-crop or general purpose tractor. By 1934, the company introduced three improved models, the FTA, KTA, and MTA; as well as a new row-crop model, the Universal JT. In 1936, the Universal Z made its debut, and in 1937 the ill-fated, vertical 2-cylinder Model YT both entered and exited the scene. In 1938, when production of the last of the original Twin City line ceased, Minneapolis-Moline introduced the U-Series.

The U-Series included: the U de Luxe (UDLX or "Comfortractor"), built from 1938 to 1940; the row-crop Universal U (UTU), the Standard U (UTS), the industrial UTI, and the high-clearance cane tractor UTC, built from 1938 to 1952 (company records indicate that the UTS remained in production into 1959); the row-crop UB, UBE and UBN, built from 1953 to 1954; and the row-crop UB Special, UBE Special and UBN Special, built from 1955 to 1957.

The Minneapolis-Moline U de Luxe or "Comfortractor" was one of the most distinctive tractors ever built. The UDLX offered the first fully-enclosed factory cab, and the first standard five-speed transmission. Equipped with windshield wipers, roll-up windows, heater, radio, rear jump seat, and top speed of 40 miles per hour, the Comfortractor could plow all day and then be driven off to town at sunset. While sales were limited to approximately 150 units over three years, the UDLX was widely promoted by the company. Among today's collectors, the "Comfortractor" is one of the most sought-after tractors of its era.

While the Universal U Tractor and Standard U Tractor were less glamorous than the UDLX, they were popular nonetheless. Offering a 4.25 x 5 inch 4-cylinder engine, in both high- and low-compression versions, the UTU was rated at 38 brake and 28 drawbar horsepower. Promoted as a 3- to 4-plow tractor, the UTS was rated at 42 brake and 39 drawbar horsepower. Both tractors were fitted with a standard 5-speed transmission. Their original styling featured a red-faced mesh type grille, which was soon replaced by a tapered, louvered grille. Other features introduced through the course of production included an LP-Gas option (1941), and live pto (1952).

In 1953, Minneapolis-Moline introduced the Model UB. Distinguished by its foot clutch, the UB was offered with gas, tractor fuel, LP-Gas, and diesel engines, and featured 12-volt ignition. Equipped with a gasoline engine, the UB was rated at 50 brake and 45 drawbar horsepower. In 1955, Minneapolis-Moline introduced the UB Special. Fitted with a flat, louvered grille, it featured power steering, live hydraulics, and interchangeable front-end assemblies.

The U-Series tractors were built for twenty years. While production peaked at 9,424 units in 1949, total production exceeded 85,000 units, making it Minneapolis-Moline's most popular series of tractors.

Aerial view of the Minneapolis-Moline Lake Street tractor plant. Minneapolis, Minnesota. *(A.H. Jensen Commercial Photography, Minneapolis)*

Cincinnati milling machine set up for milling steps in connecting rod caps. *(A.H. Jensen Commercial Photography, Minneapolis)*

Gisholt Simplimatic set up to bore, turn, and face a UT drive wheel center. *(A.H. Jensen Commercial Photography, Minneapolis)*

Model UTS (standard axle) on the Lake Street part stacking assembly line.

Lake Street plant final inspection area (foreground). Circa 1948.

Loading UTS tractors at the Lake Street plant.

MODEL U DE LUXE (UDLX)
"COMFORTRACTOR"

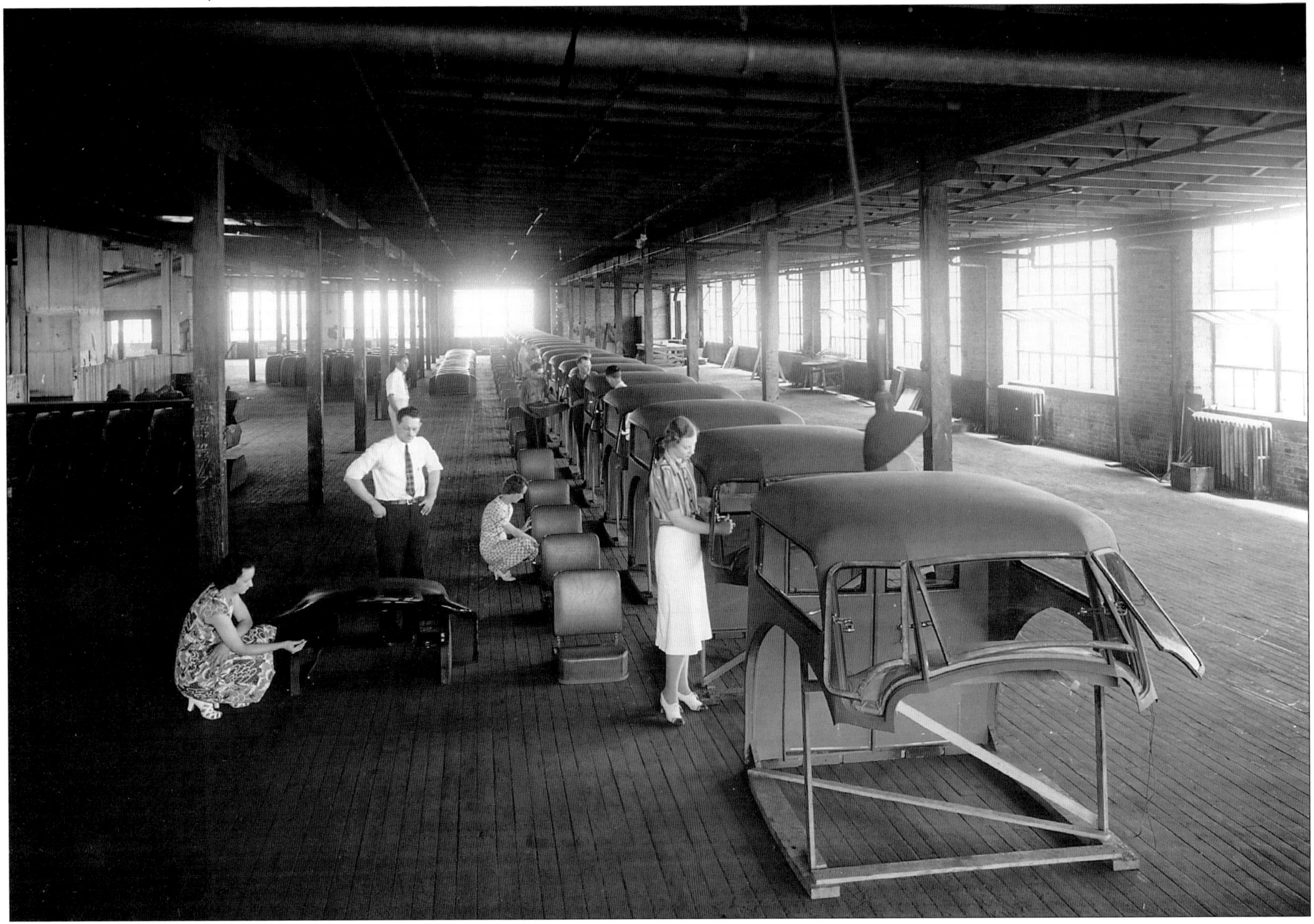

Assembly area for Model U De Luxe cabs. *(A.H. Jensen Commercial Photography, Minneapolis)*

Model U De Luxe (UDLX) or "Comfortractors" lined up in front of the Minneapolis-Moline Omaha, Nebraska offices. December 1938. *(A.H. Jensen Commercial Photography, Minneapolis)*

Comfortractor introduction at Weidner Implement Company, Freeport, Minnesota. Minnie Moline in fur coat. *(A.H. Jensen Commercial Photography, Minneapolis)*

Six Comfortractors lined up in front of the headquarter's building. Minneapolis. *(A.H. Jensen Commercial Photography, Minneapolis)*

Company salesmen line up with UDLX tractors in front of Omaha branch office. *(A.H. Jensen Commercial Photography, Minneapolis)*

Comfortractors lined up in front of the Northern Division Office ready for kick-off. *(A.H. Jensen Commercial Photography, Minneapolis)*

Comfortractors lined up in front of the Southwestern Division Headquarters, Kansas City, Kansas. December 1938. *(A.H. Jensen Commercial Photography, Minneapolis)*

Kansas Senator Arthur Capper about to enter MM Comfortractor, during Comfortractor Days in Topeka, Kansas. December 1938. *(Wolfe)*

UDLX cab interior and controls. *(A.H. Jensen Commercial Photography, Minneapolis)*

A view down the road from the Comfortractor cab. *(A.H. Jensen Commercial Photography, Minneapolis)*

UDLX steering wheel hub and gauges. *(A.H. Jensen Commercial Photography, Minneapolis)*

View of the UDLX engine compartment. *(A.H. Jensen Commercial Photography, Minneapolis)*

Minnie Moline postcard featuring UDLX Open model. *(Anderson)*

(A.H. Jensen Commercial Photography, Minneapolis)

UDLX Open and Cab models. Kansas City Division Headquarters. 1938.

Goodrich Silvertown
Goodrich Silvertown

UDLX and Model B sheller. Flagler, Colorado.

Comfortractor at highway speeds. *(A.H. Jensen Commercial Photography, Minneapolis)*

Disking with UDLX. *(A.H. Jensen Commercial Photography, Minneapolis)*

The first UDLX sold in the Midwestern Division. Yellow Jacket, Colorado. December 1938. *(A.H. Jensen Commercial Photography, Minneapolis)*

UDLX and Moline-Monitor grain drill. *(A.H. Jensen Commercial Photography, Minneapolis)*

Comfortractor and Minneapolis-Moline Huskor corn picker. *(A.H. Jensen Commercial Photography, Minneapolis)*

Model UDLX pulling Huskor corn picker and wagon. *(A.H. Jensen Commercial Photography, Minneapolis)*

Allen Tuttle, Marshalltown, Iowa, owner of the world's first Comfortractor. December 15, 1938.

The first Comfortractor sold in the Midwestern Division. Yellow Jacket, Colorado. December 1938. *(A.H. Jensen Commercial Photography, Minneapolis)*

MM Comfortractor going through the Bertroud Pass at the Great Continental Divide, elevation 11,360 feet.

 UDLX fitted with BROS snow plow. February 1939. *(A.H. Jensen Commercial Photography, Minneapolis)*

NO
LEARANCE
MINNEAPOLIS-MOLINE
MINNEAPOLIS-MOLINE

Comfortractor and snow plow. *(A.H. Jensen Commercial Photography, Minneapolis)*

UDLX and Stockland Whippet No. 2 grader. *(A.H. Jensen Commercial Photography, Minneapolis)*

STOCKLAND
STOCKLAND
WHIPPET
No 2
WHIPPET

MM
MM
MM
Firestone
Firestone

MODELS UTU AND UTS

The Visionlined Universal Tractor

"Thrift Chief of Row Crop Tractors"

The Visionlined Standard Tractor

"Power at a Touch"

Advertising slogans from 1940

Model U tractor photographed in front of the Clarkfield, Minnesota dealership.

Z, R, and U tractors on display. *(Daniel Photo)*

Two UTU tractors and plows.

The first factory-built LP-Gas Model U tractor. November 1940.

The first factory-built LP-Gas Model U tractor. November 1940.

LP-Gas UTU preparing potato seed bed. Wasco, California.
(A.H. Jensen Commercial Photography, Minneapolis)

UTS and Wheatland disk plow.

Model UTU and Wheatland disk plow.

UTU and Moline-Monitor drill.

"Quick on - Quick off" 4-row cultivator with hydraulic lift mounted on Model UTU. 1949.

Models RTU and UTU and Model 69 Harvestor.

UTU tractor and G4 Harvestor.

UTU and Harvestor.

UTU tractor and G4 Harvestor.

UTU and G-4 Harvestor.

Models RTU and UTU in threshing operation.

Model UTU and No. 69 Harvestor.

UTU and 1-row Huskor.

UTU and corn sheller.

UB AND UB SPECIAL

The UB Special Tractors

"Engineered for Bigger Operating Profits"

"Adaptable for Every Job"

Advertising slogans from 1955

MORE POWER

HIGH-TURBULENCE COMBUSTION CHAMBER

The combustion chamber used in the UB engine utilizes the advantages of both high-turbulence combustion and efficient cooling of the combustion chamber to bring you steady dependable power.

The combustion chamber causes the fuel mixture to form two air swirls, one on each side of the center of the piston, as it is compressed in the cylinder chamber. This swirling, or turbulence, results in a more complete mixing of fuel and air, more even burning, and more complete burning of the fuel.

HIGH-COMPRESSION ENGINE GIVES YOU MORE POWER

The UB engine brings you more power and greater fuel economy along with the same steady dependable performance for which all MM tractors are noted. Improved engine design and improved fuels have made it possible to increase the compression ratio of the UB engine, bringing you more power at less cost.

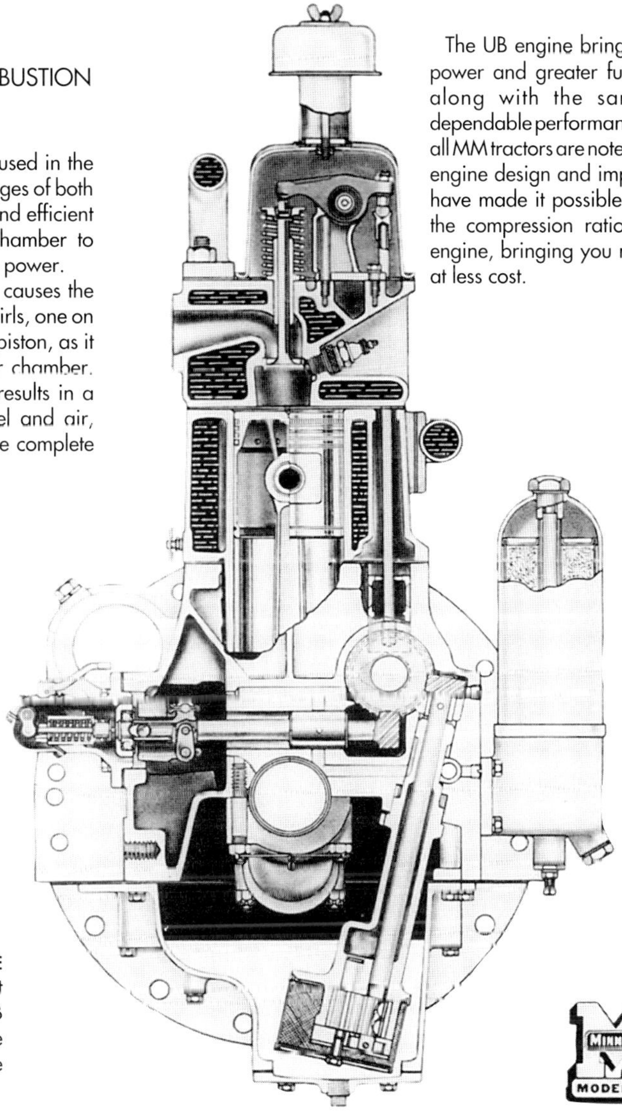

LONGER LIFE STELLITE VALVE INSERTS - stellite valve insert material is used in the new UB engine. These inserts give increased valve life and reduce valve maintenance costs.

MINNEAPOLIS-MOLINE MODERN MACHINERY

From the sales brochure *THE NEW UB TRACTORS.*

From the sales brochure *THE NEW UB TRACTORS.*

UB Special LP-Gas version.

UBE Diesel Special with extendible front axle, and Model 11B disk harrow.

UBE LP-Gas Special and Variable Drop 4-row corn planter.

UNIVERSAL
VARIABLE DROP

UBE LP-Gas Special and Model 11B disk harrow.

UB LP-Gas Special.

UBE Diesel Special and Moline-Monitor grain drill.

MINNEAPOLIS-MOLINE UB SPECIAL
DIESEL

UBE Diesel Special and rotary hoe.

UB LP-Gas Special and rotary hoe.

UBE Diesel Special with side delivery rake.

UBE Diesel Special with peg-tooth harrow.

UB LP-Gas tractor with Minneapolis-Moline 2-row cotton Harvestor.

LP-Gas UB Special with disk harrow.

LP-Gas Model UB Special and disk tiller.

UB LP-Gas Special with UO tandem disk harrow.

UB with Moline LS 300 spreader.

LP-Gas UB Special with tumblebug plow.

MM

MINNEAPOLIS-MOLINE UB SPECIAL

UBE Diesel Special and tumblebug plow.

MINNEAPOLIS-MOLINE UB SPECIAL

Diesel UBE Special Diesel and UO tandem disk harrow.

LP-Gas UBE Special and corn harvester.

Diesel UBE Special with corn picker and wagon.

Diesel UBE Special with Huskor corn picker and wagon.

INDUSTRIAL APPLICATIONS

Industrial U and sheepfoot roller.

UTI and Lull Loader operating in a gravel pit.

TRUCK
OWNERS
IULL

UTI and Lull loader.

UTI and Reese loader.

Industrial U and TAMPO roller.

UTI with shovel loader snow plow.

STUDIO ARTWORK
AND PHOTOGRAPHY

MINNEAPOLIS-MOLINE
TRACTORS & FARM MACHINERY
See Your
M-M
Dealer
TRACTORS · THRESHERS · THE HARVESTOR · FARM MACHINERY

MM

MINNEAPOLIS-MOLINE

MINNEAPOLIS-MOLINE POWER IMPLEMENT COMPANY

MINNEAPOLIS, MINNESOTA, U.S.A.

TRACTORS-POWER UNITS-FARM MACHINERY-IMPLEMENTS

Front view of Model UTC cane tractor.

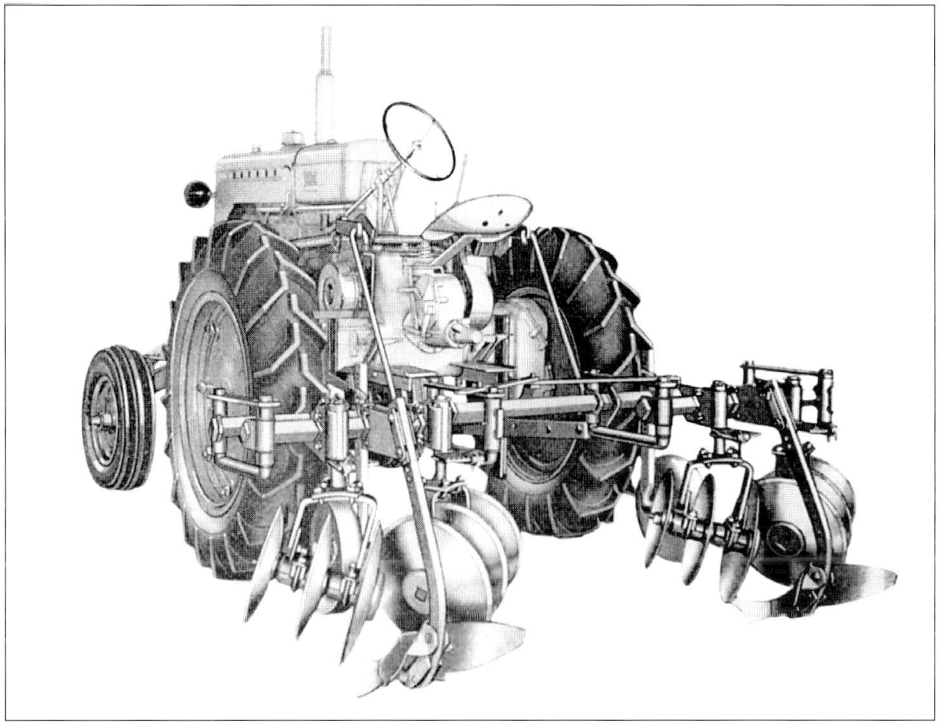

Illustration of Model UTC with cane cultivator.

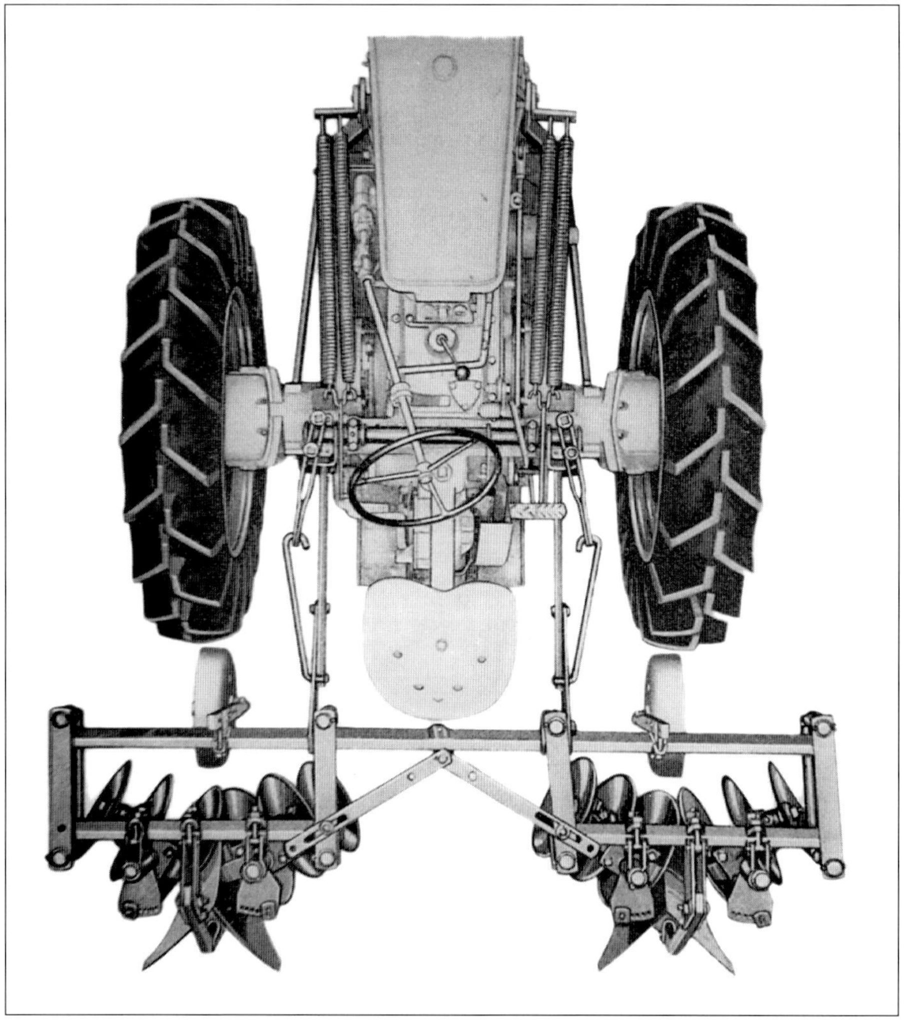

Overhead rear view of Model UTC with cane cultivator.

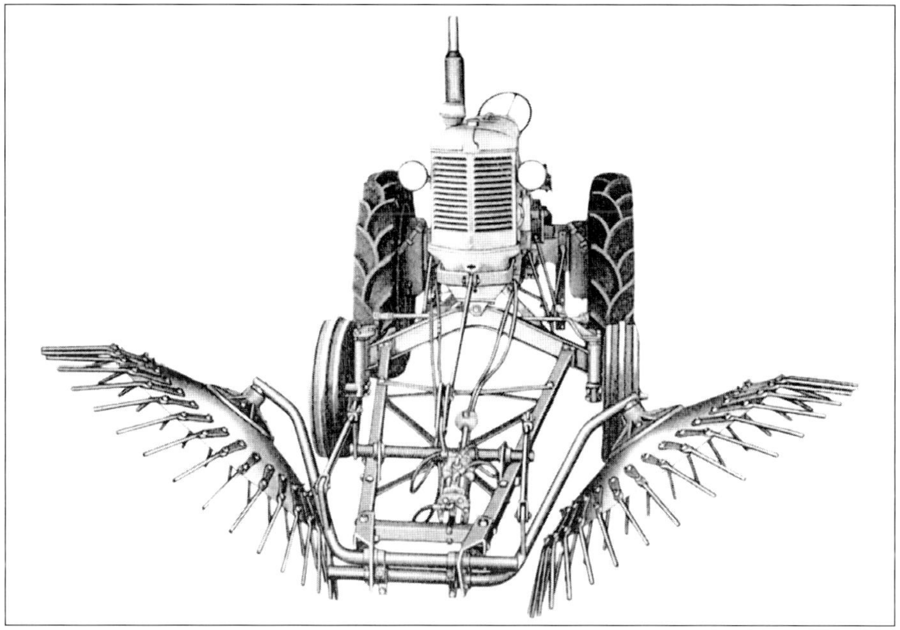

Illustration of Model UTC with cane rake.

Rear view of Model UTC cane tractor.

Left front view of Model UTC cane tractor.

LP-Gas UBE with extendible front axle.

Diesel UBE.

Close-up of Model UB front axle mounting.

Model UB extendible front axle.

Model UBN with single front tricycle wheel.

Standard Diesel UTS.

Two views of Diesel UBU Special.

MM
MINNEAPOLIS-MOLINE
UB
DIESEL

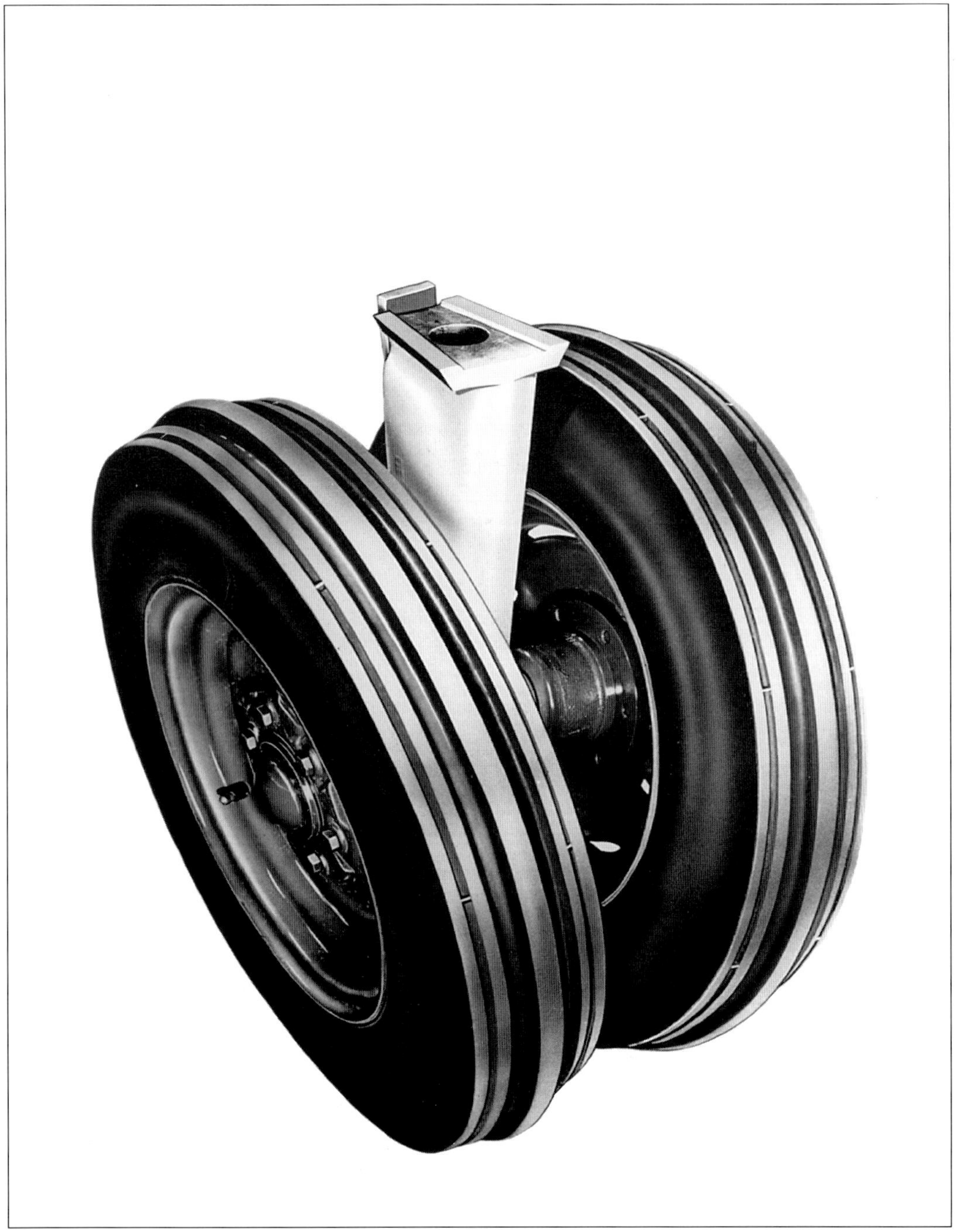

Model UB tricycle front axle.

Model UB Special gasoline engine.

The Iconografix Photo Archive Series includes:

JOHN DEERE MODEL D Photo Archive	ISBN 1-882256-00-X
JOHN DEERE MODEL B Photo Archive	ISBN 1-882256-01-8
FARMALL F-SERIES Photo Archive	ISBN 1-882256-02-6
FARMALL MODEL H Photo Archive	ISBN 1-882256-03-4
CATERPILLAR THIRTY Photo Archive	ISBN 1-882256-04-2
CATERPILLAR SIXTY Photo Archive	ISBN 1-882256-05-0
TWIN CITY TRACTOR Photo Archive	ISBN 1-882256-06-9
MINNEAPOLIS-MOLINE U-SERIES Photo Archive	ISBN 1-882256-07-7
HART-PARR Photo Archive	ISBN 1-882256-08-5
OLIVER TRACTOR Photo Archive	ISBN 1-882256-09-3
HOLT TRACTORS Photo Archive Available December 1993	ISBN 1-882256-10-7
RUSSEL GRADERS Photo Archive Available December 1993	ISBN 1-882256-11-5

The Iconografix Photo Archive Series is available from direct mail specialty book dealers and bookstores throughout the world, or can be ordered from the publisher.
For information write to:

Iconografix
P.O. Box 609
Osceola, Wisconsin 54020

or

Telephone: (715) 294-2792
(800) 289-3504 (US and Canada only)
Fax: (715) 294-3414